Lillian The Elemental Dragon With Lithium Magic

Element 3 – Magical Elementals Chemistry Fantasy

By Sybrina Durant

Lillian The Elemental Dragon With Lithium Magic

Element 3 – Magical Elementals Chemistry Fantasy

Paperback ISBN 978-1-942740-72-8

BISAC Codes:
JUV036000 - JUVENILE FICTION / Science & Technology
JUV037000 - JUVENILE FICTION / Science Fiction, Fantasy, Magic
JUV002270 JUVENILE FICTION / Fantasy / Dragons, Unicorns & Mythical Creatures

League City, Texas, United States of America

Contact Sybrina@sybrina.com.

No Metal
Lillian
The Dragon with the Lithium Tipped Tail
No Magic

Chapter 1: The Spark That Started Everything

The village of Cinderhollow sat at the edge of the Verdant Shelf, a natural limestone plateau that jutted out over the lower marshlands of MarBryn like the prow of a great stone ship. Copper-pipe streetlamps lined its cobblestone roads, hissing softly with alchemical gas that burned a warm amber each evening. Beside them, newer solar-lattice panels had been bolted to rooftops by the town's young engineers, their crystalline faces drinking in the day's light and feeding it into the humming battery-grid beneath the streets. Cinderhollow was, in every sense, a place where the old world and the new one held hands and argued politely.

It was here, seated cross-legged on the wide stone railing of the Aqueduct Bridge, that Lillian the Elemental Dragon watched the morning fog roll in off the marsh.

She was a small dragon by any standard — nose to tail-tip she measured just under six feet, no taller at the shoulder than a large deer, with a wingspan that unfolded to nearly eleven feet when fully extended. Compact and graceful, she had a hide the soft, luminous color of sea glass. Her scales caught the pale light and scattered it in gentle fragments, making her glow faintly even in fog. The ridge of delicate plates running along the crown of her head shifted through shades of lavender when she was deep in thought, and right now they were pulsing steadily. Her large, almond-shaped eyes — warm amber at the iris, edged in the faintest violet — were fixed on the horizon with an expression of careful attention. The two bracelets on her wrists — one silver-chased with tiny turquoise beads, the other lilac-dusted with a softly twinkling charm — caught the light as she turned a small journal page with one white-tipped claw. A diamond-bright charm floated lazily near the tip of her tail, orbiting it in slow, contented loops the way a moon circles a planet.

For all her size, Lillian was remarkably light on her feet — and lighter still in the air, a fact she attributed, when asked, to simply being made of the right stuff. She weighed barely forty pounds, an amount that struck most visitors as absurd for a creature with a six-foot wingspan and a hide of solid-looking scale. Bowen had weighed her once, on a dare, using a grain scale borrowed from the mill. He had checked the number three times and then refused to discuss it further.

She was writing down observations. Lillian always wrote down observations.

The eastern battery grid drew twelve percent more current overnight. No storms. No explanation on record. Third time this week.

"You're going to fall off that railing one day," said a voice from below.

Lillian didn't look up. . .or down. "I have wings, Bowen."

A stocky young man with ink-stained fingers and a leather tool belt climbed the last step of the bridge stairs, breathing harder than he probably wanted to admit. Bowen Trell was seventeen, broad-shouldered for his age and already an inch or two taller than most of the grown men in the village smithy, with a mess of dark brown curls held back by a pair of copper-framed goggles he wore more as a headband than eye protection. His hands were permanently stained with a faint gray-black smudge of graphite and machine oil that no amount of scrubbing ever fully removed, and his tool belt carried more pockets than seemed structurally reasonable, each one holding some specific implement he could produce with unsettling speed. He was MarBryn-born, Cinderhollow-raised, and the most naturally gifted electrical engineer the village had produced in a generation — a fact he was entirely aware of and mentioned perhaps too often.

He had no magic. What he had instead was a mind like a circuit board: precise, practical, and capable of solving problems that stumped people twice his age. He also had, tucked into his largest pocket, a small paper twist of honey-grain bread still warm from Marta's Bakery, which he handed up to Lillian without comment, the way he did most mornings.

"The east-side battery vault tripped its failsafe again," he said, pulling out his own notebook and flipping it open. "Third time this week."

Lillian finally looked at him. "I know. I was just writing about it." She accepted the bread with a small nod of thanks and broke off a piece, though what she actually needed after a long night of watching the grid was something else entirely — she uncorked the flask at her hip and took a slow drink of lithia water, the faintly mineral, faintly sweet spring water drawn from the limestone aquifer beneath the Shelf. It was, as far as anyone in Cinderhollow understood it, simply Lillian's preferred drink. She said it settled her thinking. Bowen had learned not to ask further questions about it, mostly because her answers involved more chemistry than he had patience for before breakfast.

"Something is drawing power out of the grid. Not using it — pulling it. Like a siphon." Bowen leaned on the railing beside her, frowning at his own notes, tearing off a piece of his own bread and eating it standing up, which was how he ate most meals when he was thinking hard about something else. "I ran the diagnostic last night. The current isn't being consumed. It's being redirected somewhere outside the grid entirely."

Lillian closed her journal. The diamond charm near her tail pulsed once — a habit it had when she sharpened her focus. "Show me."

But first, she wanted a look from above.

She stepped off the railing without ceremony, wings unfurling in a single fluid motion, and let the morning air catch beneath them. Flight, for Lillian, never felt like fighting gravity the way it clearly did for the herons that startled up out of the marsh grass below. It felt more like agreeing with it — a quiet negotiation in which her body simply weighed less than it had any right to, and the sky obliged. She called this Buoyant Lift when Sera later pestered her for a name for it, though at the time it was simply what she did every morning: bank once over the aqueduct, climb until the whole sweep of Cinderhollow's rooftops and solar-lattice panels lay beneath her like a scattered handful of coins, and glide the length of the battery-grid's underground run, feeling — more than seeing — the places where its current pulsed strong and the one place, toward the east, where it didn't.

Her sea-glass scales caught the early light as she banked, throwing small scattered flashes across the rooftops below, and for a moment two children on their way to the schoolhouse stopped to point up at her, the way they always did. She dipped one wing at them, an old habit, and came back around toward the bridge.

She landed beside Bowen with barely a sound, folding her wings in three practiced motions. "The eastern quarter," she confirmed. "I can feel it from here. Something is pulling harder than it was even yesterday."

They walked together through the waking village, past a bakery already steaming with the smell of honey-grain bread and past the open workshop of Master Vell, who was fitting a mechanical knee joint to an elderly farmer using both a hand-tool and an alchemical soldering rod in each hand simultaneously. MarBryn's blend of tradition and innovation was everywhere if you knew how to look for it.

The battery vault was housed underground beneath the town's central square — a wide, vaulted chamber lined with banks of lithium-alloy cells the size of wardrobe trunks, their indicator lights blinking in steady rhythms. Except for one entire row. Those cells were dark.

Lillian placed her palm flat against the nearest dead cell. She closed her eyes. Her mint-green scales shimmered, and a faint electric hum rose from her white-tipped claws as she read the residual current signature the way a tracker reads footprints.

Her eyes opened. The lavender plates along her crest flattened slowly, the way they did when something troubled her deeply.

"This wasn't a malfunction," she said quietly. "Someone did this deliberately."

Bowen stared at her. "Who?"

The diamond charm at her tail stopped orbiting entirely.

"Someone who knows exactly what lithium energy feels like," she said, "because they have it too."

Chapter 2: The Alchemist of Ashfen Hollow

The name came to Lillian before the sun reached midday.

She had spent the better part of the morning in the vaulted archive room above the Cinderhollow Hall of Records, her journal open beside a stack of hand-bound technical ledgers and two flickering holographic scrolls — thin sheets of alchemically treated film that displayed text in soft blue light when pressed against a reading panel. Old world and new world side by side, as always.

Bowen had gone back to the battery vault with a set of copper-thread probes and a portable current analyzer, determined to map exactly how much energy had been taken and over what duration. He worked best when left alone to think in circuits. Lillian worked best when left alone to think in patterns.

The name she found was "Solvane Criix".

He appeared in the records first as a research correspondent — a chemist-mage who had studied at the Prism Institute of Elemental Sciences in the city of Dralkethorn, MarBryn's sprawling northern capital. Twenty years ago, Solvane Criix had been considered a prodigy in the field of electrochemical magic: the rare discipline that merged alchemical battery science with elemental spellwork. His early papers, recorded in the archive on original parchment, were brilliant.

Then the records changed tone.

A disciplinary note, dated twelve years back, described unauthorized experiments using mass-scale energy siphoning arrays. A second entry, barely a year later, recorded his expulsion from the Institute for "destabilizing elemental resonance fields across the northern shelf in pursuit of unethical amplification research."

After that — nothing. No address. No correspondence. No record of death.

Lillian traced one white-tipped claw down the last entry and sat back in her chair, the lavender plates along her crest rising slowly with the particular rhythm they had when pieces of a puzzle locked together.

"*He's still out there. And he hasn't stopped*."

She tucked her journal under her arm and was halfway down the archive stairs when she nearly collided with someone coming up.

The figure stumbled back, catching the railing with both hands — and Lillian caught a glimpse of bright copper-red hair, a long coat the color of storm clouds, and a pair of remarkable eyes the precise green of oxidized bronze.

"Watch where you're — " the stranger started, then stopped. Her eyes went wide. "You're the elemental. The lithium dragon."

Lillian looked at her carefully. "You have the advantage of me."

"Sera Vance." The young woman straightened her coat with the crisp efficiency of someone accustomed to composing herself quickly. She was perhaps twenty, of middling height, lean and sharp-featured, with the kind of quick, precise movements that suggested she had spent a great deal of her life around instruments that broke if handled carelessly. Her copper-red hair was cropped short and practical, pinned back from her face with what appeared to be an actual laboratory clip rather than any ornament, and a scatter of small burn scars across the back of her left hand told their own story about a childhood spent too close to unstable compounds. She carried a satchel across one shoulder bristling with vials, folded instruments, and what appeared to be a compact alchemical spectrometer. "I'm a field chemist. Independent. I've been tracking anomalous energy draws across the Verdant Shelf for the last six weeks, and every trace leads here." She paused. "To your battery grid."

"Come with me," Lillian said.

They found Bowen in the square, already packed and irritated, his current analyzer wrapped in cloth and tucked into his pack. He greeted Sera with a polite nod and the immediate suspicion he reserved for anyone carrying equipment he hadn't built himself.

"Her spectrometer is better than your analyzer," Lillian told him.

Bowen looked at the device in Sera's satchel. His mouth tightened. "It is not."

"It is," Sera confirmed without apology.

Lillian almost smiled.

She laid out what she had found in the archive. Sera confirmed it from her own six weeks of fieldwork: the siphon signatures traced a northeast line, moving toward the deep fog-country beyond the Shelf's edge — a region of dense alchemical fog and unstable terrain known as Ashfen Hollow. Few maps included it. Fewer people chose to go there.

"Criix has been building something," Sera said, spreading a hand-drawn chart across the square's fountain edge. "Each draw from the battery grids across the Shelf has been larger than the last. He's charging something. Something big."

"How big?" Bowen asked.

Sera hesitated for just a moment. "Based on the cumulative energy volume? Big enough to destabilize every elemental grid on the Shelf. Maybe farther."

The silence that followed was the kind that settles when a problem stops being interesting and starts being serious.

Lillian looked north, toward the fog-line where Ashfen Hollow swallowed the horizon. Her bracelets caught the noon light — a brief, quiet shimmer of silver and lilac.

"Then we go," she said simply.

Bowen pulled his goggles down over his eyes with the resigned expression of someone who had always known this was coming. "I packed last night," he admitted.

Sera was already folding her chart, and — with the same brisk efficiency — tucking a twist of dried, sour-pickled root vegetables into her coat pocket, the traveling snack she kept on hand at all times and rarely offered to share. "I packed three days ago."

Chapter 3: The Road Through Prism Bog

The path northeast from Cinderhollow's plateau descended steeply through switchback trails cut into the limestone face, bottoming out into a wide transitional zone between the Verdant Shelf's organized farmland and the wild, untamed terrain that MarBryn's cartographers had historically labeled with the elegant understatement: -*Here, conditions vary*.-

Prism Bog was the first major obstacle — a vast wetland basin where naturally occurring alchemical minerals had leached into the standing water over centuries, producing a phenomenon that was both beautiful and deeply hazardous. The bog's surface shimmered with a prismatic iridescence, the water cycling through every color in the visible spectrum as trace elemental compounds reacted continuously with the fog and the light. At night, it glowed. During the day, the shifting colors could confuse the eye so completely that travelers walked in circles for hours without realizing it.

"Stay in single file," Lillian said, stepping onto the first solid tussock. "Don't look at the water directly. Look at the path."

"What happens if you look at the water?" Bowen asked, looking at the water.

"Your spatial orientation unravels. You think you're walking forward and you've turned one hundred and eighty degrees."

He stopped looking at the water.

Before they committed to the bog itself, Lillian spread her wings and took to the air for a wide, looping scout above the tree line — Buoyant Lift carrying her up in a few unhurried wingbeats until Bowen and Sera were two small shapes on the trail below. From this height the bog looked less like water and more like a vast, slow-shifting opal, its colors bending and pooling in patterns that made no sense from the ground but resolved, from above, into a rough map: a firmer causeway of pale stone running roughly northeast, and around it, wide unstable stretches where the color moved too quickly to be anything but deep, waiting mud. She circled twice, marking the safe line in her mind the way she marked everything, then folded her wings and dropped in a smooth, controlled spiral that ended with her claws settling silently back onto the tussock beside Bowen, who had, by that point, stopped being surprised by this kind of thing.

"There's a clean line about ten degrees east of where we're standing," she said. "I'll lead."

They moved through the bog in careful procession, Lillian leading. Her tail constantly swished back and forth, the diamonds circling the tip generating a soft, steady pulse of electrokinetic energy — a compass she had learned to produce intuitively, reading the electromagnetic field beneath the bog floor to distinguish solid ground from soft mud hidden beneath the shimmering surface. Tiny arcs of pale blue-white light danced across the air as she worked, lighting the path ahead in brief, beautiful flickers.

"That's how you navigate?" Sera asked from behind her, scribbling notes as she walked. "You're reading subsurface mineral conductivity."

"Essentially," Lillian said. "The compressed lithium-slate deposits under the solid ground paths conduct differently than the saturated mud. I can feel the difference."

Sera wrote something very quickly and underlined it twice.

They were two hours into the bog when Bowen stopped and held up a fist — a signal he had apparently decided to use without consulting anyone, but which communicated "stop" effectively enough.

"Something's ahead," he said quietly. "Look at the gas-pipe marker."

At the edge of the path, one of the old iron marker posts — part of a now-defunct alchemical fuel line that had once run through the region — had been bent at a sharp angle and draped with a length of what appeared to be copper wire, wound in a very specific coil pattern.

Lillian crouched and examined it without touching. Her scales rippled faintly. "It's a detection snare. Electrochemical. It transmits a signal when disrupted." She looked up. "Someone wanted to know if anyone was following this path."

"Criix?" Sera asked.

"His signature. It matches the residual current I read in the battery vault."

"Can you neutralize it?" Bowen asked.

In answer, Lillian pressed two claws lightly to the coiled wire and drew the stored current out of it with smooth, practiced ease — a technique she had started calling Voltaic Draw, after Sera pointed out that "the thing you just did with your claws" needed a name for her notes — absorbing the energy into herself the way a conductor grounds a charge. The wire went dark. The faint hum it had emitted stopped.

"Done," she said.

They continued. Around midday they paused on a wide, flat stone that jutted above the fog line, offering a brief, clear view of the landscape ahead. The Ashfen began roughly a kilometer north — they could see where the prismatic bog colors faded into a heavy, gray-green chemical fog that moved differently from weather fog, slow and deliberate, as if thinking.

Sitting on the stone, waiting for them with her arms crossed and an expression of practiced impatience, was a creature none of them had expected.

She was small — barely waist-high to Bowen, which put her somewhere around two and a half feet standing — with iridescent wings folded tight against a compact body covered in overlapping scales of deep teal and gold. Her eyes were the bright, vivid yellow of sulfur crystal, sharp and quick, and she carried a short staff tipped with a spinning globe of what appeared to be compressed atmospheric gas sealed in alchemical glass. Her coat, a patchwork of deep greens too large for her frame, was cinched at the waist with a belt of small pouches that clinked faintly whenever she moved, and would later turn out to contain — among other things — a truly improbable quantity of crystallized honey candy, which she ate constantly and shared with absolutely no one.

"You're late," she announced. "I've been sitting here since dawn."

"We weren't aware of an appointment," Lillian said carefully.

The creature unfolded from the rock with fluid, effortless grace. "My name is Zix. I'm a

Compass Sprite — specifically, I'm the only one who knows a safe path through Ashfen Hollow. And I'm told you need one." Her sulfur-yellow eyes moved to each of them in turn. "I was sent by the Elemental Council of Dralkethorn. They know what Criix is building. They're worried." She paused. "They should be."

Bowen looked at Lillian. Lillian looked at Zix.

"What is he building?" Lillian asked.

Zix's wings flicked open once — a gesture that read, unmistakably, as finally, the right question.

"A resonance engine," Zix said. "One designed to run on pure elemental lithium energy." Her gaze settled on Lillian with steady, serious weight. "On your energy, specifically. And he's nearly finished charging it."

Chapter 4: What Zix Knew

They made camp on the high stone shelf above the bog's edge as the afternoon light thickened into gold. Bowen assembled a compact thermal-coil stove from his pack — a clever folding design of his own invention that burned alchemical fuel pellets cleanly and quietly — while Sera laid her charts flat across the rock surface and began updating them with Zix's information using a stylus that recorded in two inks simultaneously: standard graphite and a faintly luminescent trace compound that would show up on her spectrometer when she reviewed the charts later.

Zix sat on her preferred elevated position — the top of Bowen's pack, which he bore with extraordinary tolerance — and talked.

Dinner, when it came together, told you almost everything you needed to know about the four of them. Bowen cooked, because Bowen always cooked when there was a stove within reach and an audience within earshot — a thick trail stew of root vegetables, dried mushroom, and smoked sausage that he seasoned with more confidence than precision, ladled into four battered tin bowls, and ate standing up despite there being perfectly good rocks to sit on. Sera ate hers methodically, seated with her charts still spread across her knees, alternating bites of stew with sips from a small flask of her own strong, bitter mineral tea and, when she thought no one was watching, the sour-pickled vegetables from her coat pocket, which she still did not offer to share. Zix declined the stew entirely in favor of three pieces of crystallized honey candy and a fourth she pretended to save for later and then also ate, washing it down with nothing at all, which Bowen found deeply unsettling to watch.

Lillian took a small bowl more out of courtesy than appetite — elementals of her kind drew most of their true sustenance from mineral-rich water and the ambient trace energy of the world around them rather than food in the way the others understood it — but she ate a few spoonfuls anyway, and refilled her flask of lithia water from a clean spring trickling off the rock face, murmuring something about the local mineral content being "unusually good" that made Sera immediately want to test it.

The Elemental Council of Dralkethorn, she explained, was the governing body responsible for monitoring elemental magic use across all of MarBryn. It was composed of twelve senior elemental practitioners, one for each of the primary elements known to the magical world. The Council had been watching Solvane Criix's movements for two years, ever since their sensors detected unusual resonance disturbances in the northern territories.

"Criix was expelled from the Prism Institute," Sera said. "I found that much."

"Expelled and embittered," Zix confirmed, her sulfur eyes sharp. "He believed the Institute was suppressing his research. He argued that lithium-based elemental energy was the most potent and stable force in the known magical spectrum — capable of powering systems far beyond what any other element could sustain. He wasn't entirely wrong about the science."

"But?" Lillian prompted.

"But he designed his resonance engine to take that energy by force rather than by study. It can't generate lithium energy on its own — no machine can, because lithium magic is intrinsically biological. It lives in elemental beings." Zix's gaze went to Lillian without hesitation. "In you. The engine requires a living elemental lithium source, connected to it directly, to run at full capacity."

The alchemical stove crackled quietly in the silence that followed.

Bowen spoke first. "He's building a machine that runs on Lillian."

"On her energy, yes. A continuous forced extraction." Zix's tone was matter-of-fact but not unkind. "The preliminary siphoning from the battery grids across the Shelf has been calibration work — tuning the engine to recognize and accept lithium resonance. Once calibrated, the grid power won't be enough. He needs the source."

Lillian had been listening with the still, focused expression she wore when processing information she did not like. Her lavender crest-plates lay flat. The diamond charm at her tail was motionless.

"What does the engine do?" she asked. "Once powered."

Zix hesitated — a small, telling pause. "It amplifies and redirects elemental energy on a continental scale. Criix intends to use it to seize control of every elemental grid in MarBryn. Every battery system, every alchemical power line, every elemental circuit in the continent — all of it answering to him." She let that settle. "He would effectively own MarBryn's energy infrastructure. New tech, old tech, all of it. Whoever controls the energy controls everything else."

Sera set her stylus down. "That's not a scientific project. That's a coup."

"Disguised as a scientific project," Bowen said. "For years. He's been planning this for years."

Lillian looked at the fog bank to the north. The Ashfen moved in slow, chemical eddies, greenish-gray and heavy, occasionally lit from within by a brief bioluminescent flicker from whatever lived in its depths.

"Tell me about the engine's weak points," she said.

Zix almost smiled — the expression tightened her narrow face in a way that managed to look both pleased and fierce simultaneously. "That is exactly what I was hoping you'd ask." She hopped down from Bowen's pack, spread her wings, and hovered at chart height. "It has three primary components. A collection array, a storage core, and a projection matrix. Destroy any one of the three and the engine fails. But Criix knows this, so all three are protected."

"By what?" Bowen asked.

"By his associates," Zix said. "He has recruited. Specialists. People who share his grievance against the Council, or who simply want to be on the winning side of a continental power shift." She folded her wings and landed on the chart's edge. "Expect resistance before you reach the engine."

Lillian picked up her journal, opened to a clean page, and began writing in her precise, compact style.

Engine structure: three components. Collection array, storage core, projection matrix. Approach from the northeast to avoid primary detection field. Criix is not working alone.

She looked up. "We move at first light."

Bowen pulled a second fuel pellet from his pack and fed it to the stove, then scraped the last of the stew from the pot directly into his own bowl for a second helping, on the grounds that cold stew was a crime he refused to commit twice in one trip. "I'll upgrade the spectrometer to detect the engine's frequency overnight," he told Sera.

Sera looked at him with something approaching respect, though she managed to keep it off her face almost entirely. "I'll help."

"I know you will," he said, and handed her a calibration wrench.

Chapter 5: Ashfen Hollow and the Fog Wolves

They entered Ashfen Hollow at dawn, with Zix leading and the fog closing around them within twenty paces of the boundary line.

The Hollow was unlike any terrain Lillian had traveled. The ground was soft and uneven — compressed layers of ancient alchemical sediment that gave slightly underfoot, releasing a faint sulfuric tang with each step. Strange formations rose from the fog at intervals: towers of crystallized mineral deposit in burnt orange and pale yellow, some as tall as houses, their facets catching and scattering what little light penetrated the fog canopy. Older iron pipework, rusted to deep reddish-brown, ran in disorganized lines between the formations like the skeletal remains of some long-dead industrial system. Newer conduit — thin, polymer-coated cable — ran alongside it in places, evidence that Criix had been running his own infrastructure into the Hollow.

"He's laid cable all the way from the Shelf," Bowen muttered, crouching to examine the conduit. "This is high-grade transmission line. Expensive."

"He's been funded," Sera said. "Someone with resources has been backing this."

"The Council suspects a merchant faction from the Eastern Ports," Zix said without slowing her pace. "Trade interests who would benefit from controlling MarBryn's energy supply. We don't have names yet."

They moved deeper. The fog thickened. Zix navigated with confidence, her staff's compressed-gas globe spinning faster when she was on course, slower when she deviated — a navigational instrument as elegant as it was unusual.

The attack came without warning.

One moment the path was clear. The next, shapes were materializing from the fog — four of them, low and swift, with bodies that seemed to be formed of compressed chemical vapor held in vaguely lupine shape by crackling internal energies. Their eyes burned amber. Their paws left faint scorch marks on the sediment.

"Fog Wolves," Zix snapped. "Criix's sentinels. Don't let them touch you — their contact charge is nasty."

The first wolf lunged at Bowen. Lillian's paw shot out and a bolt of lightning cracked from her clawtips, catching the creature mid-leap. It detonated in a burst of dispersed vapor and residual energy — not dead, exactly, since these were constructs rather than living animals, but disrupted enough to dissolve momentarily.

"*Three remaining.*"

Sera had her spectrometer up, scanning rapidly. "They're sustained by a carrier frequency — there's an emitter nearby. Destroy the emitter and they all collapse."

"Where?" Bowen demanded, back-stepping from a snapping wolf.

"North-northeast, about forty meters. Behind the orange crystal formation."

Lillian reached out with her telekinesis — Lattice Grip, she called it later, after Sera asked what to write down, because it felt to her like reaching into the orderly, invisible lattice that held any solid object together and simply asking it to move — a focused concentration, a directed pull of will, and lifted the largest loose crystal fragment from the ground beside her. It was heavy, perhaps forty kilograms, nearly the weight of Lillian herself, and she held it aloft for exactly one second before hurling it at the crystal formation with the full force of her mental grip. The formation shattered. A small device hidden within it — a squat, cylinder of copper and polymer — sparked violently and went dark.

The fog wolves unwound like smoke in wind, their amber eyes fading, their shapes dispersing into the Hollow's general fog with a long, diminishing hiss.

Silence.

Bowen exhaled slowly. "Right. So those exist."

"There will be more before we reach the engine site," Zix said, already moving again.

"Can we bypass the emitter network instead of triggering it?" Lillian asked.

Sera was already analyzing her scan data. "If I can map the emitter frequency, yes. I can program the spectrometer to broadcast a suppression signal that masks our approach. It would take about an hour."

"Do it while we move," Lillian said.

Sera nodded, walking and working simultaneously, her copper-red hair catching brief glints of mineral-light through the fog as she adjusted her instrument.

It was during this hour of careful forward movement that they encountered a man.

He was sitting against a rusted pipe junction with one hand pressed to his ribs and a look of profound annoyance on his face. He was tall, gaunt, and older than the others — perhaps fifty, with close-cropped silver hair, a deeply lined but composed face, and dark eyes behind a pair of wire-framed spectacles that had been repaired, at some point, with a length of the same copper wire Bowen used for his own equipment. He wore the coat and insignia of a Dralkethorn Council field agent, though the coat had clearly seen better decades, and a battered field ration tin, empty, stuck out of one pocket — the last of a month's worth of dry biscuit and preserved meat that he had, by his own later admission, been rationing so carefully he'd started dreaming about proper meals.

"Agent Davan Marsh," he said, by way of greeting. "Council Intelligence. I've been tracking Criix for fourteen months." He grimaced and shifted his hand, revealing a spreading bruise along his ribs. "His wolves found me an hour ago. I've been waiting for someone competent to come along."

Lillian crouched beside him. She placed the palm of her paw gently against his ribs and closed her eyes. Warm energy flowed from her — a calm, golden current distinct from the sharp blue-white of her electrokinesis, a technique she thought of as a Buffer Pulse, the same steadying, stabilizing principle that made real-world lithium carbonate useful in medicine, applied here to bruised tissue instead of a troubled mind. She felt the cracked rib, the bruised tissue, and worked carefully, accelerating the body's repair with steady, methodical focus. Davan's expression changed from pained to surprised to quiet relief within the span of about a minute.

"That," he said carefully, "is remarkable."

"What do you know about the engine's location?" Lillian asked.

He straightened, testing his ribs with evident approval, and — perhaps out of pure habit after a month alone — immediately began looking around for something to eat, until Bowen silently handed him a strip of dried apple from his own pack, which Davan accepted with the gravity of a man being handed a medal. "I know exactly where it is. And I know something none of you do." He looked at each of them in turn. "Criix has a secondary fail-safe. If the engine is destroyed without being properly shut down first, it releases all stored energy in a single detonation." He paused. "The blast radius would take out everything within a half-kilometer."

Zix's wings snapped open. Bowen stared. Sera went very still.

Lillian considered this for exactly four seconds. "Then we shut it down first," she said. "Tell me everything you know."

Chapter 6: The Architecture of the Enemy

Davan Marsh was, as advertised, comprehensively informed.

Seated on a collapsed crystal formation in the fog while Sera finished calibrating the emitter-suppression signal, he walked them through everything Council Intelligence had assembled over fourteen months of covert observation. He had a sharp, methodical mind and the particular gift of intelligence professionals for presenting complex information without unnecessary ornamentation. Up close, in better light than the pipe junction had offered, he looked every one of his fifty-odd years and then some — the fog-country had worn on him visibly, hollowing his cheeks and deepening the lines around his eyes — but his hands, Lillian noticed, were perfectly steady the entire time he spoke, the hands of a man who had learned long ago not to let his body betray his nerves.

Solvane Criix had established his base of operations inside a place called the Sunken Cradle — a natural depression in Ashfen Hollow's center, roughly three hundred meters across, where centuries of mineral deposit had created a bowl-shaped basin with near-perfect acoustic and electromagnetic properties. The basin walls reflected and focused energy with extraordinary efficiency, which was precisely why Criix had chosen it.

Within the Cradle, he had constructed the resonance engine over a period of three years, using a workforce of both hired laborers and magically compelled constructs. The collection array — a vast web of copper and alchemically treated silver wire stretched between thirty-meter lattice towers — covered the entire basin rim. The storage core sat at the Cradle's center: a structure roughly the size of a large house, built from reinforced lithium-alloy plating and designed to hold accumulated energy in stable suspension. The projection matrix was housed inside the core — a rotating assembly of lens-like crystalline discs that would, when activated, focus the stored energy into a directed continental broadcast.

"The shutdown sequence," Davan continued, "requires physical access to the core's primary control panel. It's a chemical-key system — old tech, deliberately. No electrical signal can trigger it remotely because the system is galvanically isolated from everything else."

"What's the chemical key?" Sera asked.

"A specific compound introduced into the panel's intake port. I have the formula." Davan produced a folded card from his coat's inner pocket, written in dense notation. Sera took it, scanned it, and immediately began checking her supply vials.

"I have most of this," she said. "Not all. I'm missing stabilized lithium carbonate."

She looked at Lillian.

Lillian held out her paw. "Give me your smallest empty vial."

Sera handed it over. Lillian turned her attention inward, drawing on the elemental energy that lived at her core — not the sharp electrical variety but the deeper chemical resonance that connected her to lithium itself. She had never bothered naming this particular technique before, since she rarely used it in front of anyone, but Sera pressed her for a term to write in the official record, and Lillian, after a moment's thought, offered "Carbonate Bloom" — because that was, she said, more or less what it felt like: a slow, deliberate flowering of raw elemental energy into something with a fixed chemical shape. She let a small measure of that energy condense in her palm, coaxing it into compound form through the slow, meditative process her elemental nature made possible. A few crystals of pale white material formed in the vial. Not large-scale production, but enough.

Bowen watched this with the expression he wore when encountered with something his engineering training had no category for. "You just... made lithium carbonate."

"A small amount," Lillian said. "And it costs effort. Don't ask for more unless necessary."

Sera sealed the vial with practiced efficiency. "The compound will be ready to synthesize into the key mixture in about twenty minutes."

"Then we use that time to plan the approach," Lillian said.

They gathered around Davan's hand-sketched map of the Cradle, weighted at its corners by four mineral chunks Zix had produced from somewhere in her coat — a coat which, Lillian had observed, contained a great deal more than its exterior volume should reasonably permit, including, it turned out, an entirely separate and apparently bottomless reserve of honey candy that Zix ate two more pieces of during the planning session alone.

The Cradle had three entry points: a wide main approach from the south, a narrow service path from the west used by Criix's workers, and a third route — partially collapsed, marked on Davan's map with a cautious asterisk — that ran through a tunnel in the eastern crater wall.

"The south approach is guarded," Davan said. "I counted at least eight of his associates during my observation, plus fog wolf emitters at every twenty-meter interval."

"West service path?" Bowen asked.

"Monitored by chemical sensors. Automated. They'd detect Lillian's elemental signature immediately."

"The east tunnel, then," Lillian said.

"It's structurally questionable," Davan admitted.

"But not watched," Zix said. "Because no one sensible would use it."

Bowen and Sera exchanged a glance that contained an entire wordless conversation.

"Accessibility is a reasonable trade for surprise," Lillian said. She looked at each of them. "We go in through the east tunnel. Bowen, I need you to assess structural integrity as we move — if it starts to fail, tell me immediately. Sera, you're carrying the shutdown compound and you get to the control panel. Zix, you navigate." She turned to Davan. "You know the layout better than any of us. Stay close to me."

"With pleasure," Davan said drily. "I've been alone in this fog for a month. I'd welcome the company."

"One more thing," Lillian said. Her voice was level, but the lavender plates along her crest had risen slightly — the quiet signal that she was saying something important. "Criix was brilliant once. He may have believed, and may still believe, that what he's doing is right. That does not change what we have to do. But if there is an opportunity to resolve this without permanent harm to anyone, we take it."

Davan looked at her for a moment. "You'd extend that to him?"

"Particularly to him," she said. "Destroying the engine matters. Destroying people is a last resort."

The group was quiet for a moment with the weight of that.

Then Sera finished sealing the last component of the shutdown compound into a reinforced carry-case, Bowen checked his tool belt for the third time, and Zix's staff began to spin.

"East tunnel," Zix confirmed. "Follow closely. The floor does something unusual around the halfway point, and I need you to trust me when I tell you to jump."

"Jump?" Bowen said.

But she was already moving.

Chapter 7: The Battle at the Cradle Rim

The east tunnel was exactly as structurally questionable as advertised.

It ran through the crater wall at a descending angle, its walls formed of compressed sediment reinforced — barely — by iron struts that had been placed at some point in the distant past and had been rusting conscientiously ever since. Bowen moved with his hand on the wall, checking pressure and flex, announcing his findings in clipped, quiet updates. "This section is fine. That strut is carrying too much — step wide of it. The ceiling is good through here, worse ahead."

The unusual thing at the halfway point turned out to be a section of floor that had partially dissolved through chemical reaction with the underlying mineral layer, leaving a gap of roughly two meters over a drop of perhaps four. Zix cleared it without effort on her wings. Bowen cleared it on a running jump. Sera cleared it with economical efficiency. Davan cleared it with considerably more determination than grace and a suppressed exclamation that he later declined to discuss.

Lillian simply stepped off the edge and let Buoyant Lift do the rest — a half-second of true falling before her wings caught the air and her weight, already unnaturally slight for a creature her size, seemed to forget gravity entirely. She crossed the gap in a slow, unhurried glide, touched down on the far side without so much as a scuff of claw against stone, and didn't bother mentioning that she could have simply flown the whole tunnel from the outside if the ceiling height had allowed it.

Then they were through the tunnel and at the inner wall of the Sunken Cradle, and for a moment all of them simply stared.

The resonance engine was enormous. The storage core alone rose twenty meters from the Cradle floor, its lithium-alloy skin gleaming with the deep silver-pink of metallic lithium, catching the strange light that filtered through the hollow's permanent chemical fog canopy. The collection array towered above it, its web of copper and silver wire vibrating with a low, pervasive harmonic that Lillian felt in her chest — a pulling sensation, like a current moving toward a drain.

"It's already drawing energy," she said. "The calibration is done. It's running on stored charge."

"How long before it's at operational capacity?" Sera asked.

Davan studied the core's indicator lights — a ring of graduated alchemical gauges visible even from their position at the tunnel mouth. "Based on the charge rate I observed last week?" He calculated briefly. "Six hours. Perhaps less."

"Then we don't have time for a careful approach," Bowen said.

"We don't," Lillian agreed. "But we need Sera at that control panel before anything else happens. The shutdown compound has to be introduced before we touch anything else or the fail-safe triggers."

She scanned the Cradle floor below. Eight figures moved between the engine's support structures — Criix's associates, as Davan had described. They were equipped with a mixture of old and new tech: crossbows fitted with electrochemical charge bolts, polymer riot shields crackling with stored static energy, and at least two individuals carrying alchemical projectors — heavy shoulder-mounted devices that could fire concentrated acid streams at range. Serious equipment, seriously carried.

And positioned between the tunnel mouth and the core's base were six fog wolf emitters, their amber indicator lights pulsing in steady rhythm.

Lillian turned to her companions. "Sera and Davan go for the control panel — southeast face of the core, Davan said. Bowen, I need those emitters down. Zix, can you get Sera to the panel without crossing the open floor?"

Zix's wings spread. "There's a service gantry running along the inner crater wall. It connects to the core's upper level. I spotted it coming in." She paused. "It's about thirty meters up."

"I can manage thirty meters," Sera said, adjusting her carry-case strap.

"I'll carry the case," Davan said. "You'll need both hands for climbing."

Sera considered this for precisely one second and handed it over.

"I'll create the opening," Lillian said. "When I move, everyone moves."

She stepped to the tunnel mouth, took one slow breath, and then stepped out into the Cradle — but not on the ground. She launched instead, wings snapping open, climbing hard in a tight spiral that put her fifteen meters above the Cradle floor before anyone below had finished turning to look. From up there she could see the whole battlefield laid out at once: the eight guards, the six emitters, the gantry Zix had marked, the open door at the storage core's base that no one else had noticed yet.

The first guard saw her immediately — a broad figure with a static shield who pointed and shouted something that brought every head in the basin turning. Lillian folded her wings into a steep dive, leveling out barely three meters above the ground, and let the electrokinesis run free — not the precise, targeted bolt of an Ion Lance this time, but what she had started calling a Cascade Discharge: a wide-arc release that jumped from target to target the way a redox reaction cascades through a reactive series, one triggered discharge setting off the next. The current crackled across the open air in a sheet of blue-white lightning, striking all six fog wolf emitters simultaneously. The devices detonated in a chain of sharp, percussive bursts. The fog wolves that had begun to form from them dissolved instantly, their half-materialized shapes unraveling into wisps of spent vapor.

The Cradle erupted into motion.

Three guards charged toward her with crossbows raised. Lillian swept her tail laterally and her telekinesis caught all three bolts mid-flight, stopping them dead in the air for a fraction of a second before redirecting them harmlessly into the crater wall. She then seized a section of loose iron strut from the ground with the same mental grip and swung it in a wide arc, scattering the approaching guards like pins.

To her left, Bowen had engaged two guards near the collection array's base. He had no magic — what he had was a length of copper-thread cable from his tool belt that he'd connected at one end to an iron spike driven into the sediment and at the other to the array's nearest lattice tower. The static charge bleeding off the tower ran down the cable and across the ground between him and the two guards. Neither had noticed until they stepped into the field and their boots began sparking. They retreated fast, and Bowen pushed forward with a handheld arc-suppressor — a device of his own construction — that sent directed bursts of electromagnetic interference into their equipment, disabling the acid projector carried by the nearer of the two with a fizzing crack of shorted components.

"Engineering," he said to no one in particular, with satisfaction.

Zix had Sera and Davan moving along the crater wall at speed, her staff spinning rapidly as she led them to the gantry access point. Two guards who moved to intercept were met with blasts from Zix's staff — compressed atmospheric bursts that hit like a strong wind and carried enough kinetic force to knock a grown man sideways off his feet.

But three guards remained unengaged, and they were smart. They had stopped advancing. They had fanned out. And they were aiming at Sera's group on the gantry.

Lillian saw the threat before the bolts were fired. She dropped altitude fast, wings cupping the air to slow her fall at the last possible moment, and positioned the tip of her tail to project a sustained energy barrier between the shooters and the gantry — what she would later call an Ionic Ward, a name Sera insisted on for her notes, drawn from the same principle that lets lithium ions lock into a tight, stable lattice: a broad, shimmering plane of concentrated lithium energy that caught the charge bolts as they struck and absorbed them in bursts of violet-white light. Holding the barrier and maintaining combat awareness simultaneously pulled hard on her focus. She felt the effort in her chest, a deep interior pressure.

"Hold it. Just until Sera reaches the panel."

"Lillian!" Davan's voice rang from above. "We're at the panel!"

She let the barrier drop the moment Sera's hands went to the control port, and redirected everything she had left into a single, overwhelming energy pulse projected outward from her body in all directions — a Full Discharge, the same principle by which a battery can release its entire stored charge in one uncontrolled burst, except that Lillian's version was anything but uncontrolled — not targeted, not destructive, but a massive electromagnetic disruption wave that washed across every piece of active equipment in the Cradle simultaneously. Charge bolts misfired. Static shields shorted out. Two acid projectors vented harmlessly into the crater floor.

In the stunned silence that followed, the remaining guards looked at their dead equipment, looked at the small mint-green dragon standing in the middle of the Cradle floor with her scales blazing in arcs of blue-white light and her crest-plates fully raised, and made a series of rapid individual decisions about how much they were being paid.

Four of them ran. The remaining four sat down on the crater floor with their hands visible.

Bowen, mildly surprised, looked at the seated guards and back at Lillian, who had by then settled to the ground, folding her wings with the particular care she used when she was more tired than she wanted to admit. "That works too."

From the gantry above, a sharp hiss followed by a resonant, mechanical clunk announced the shutdown compound engaging the control panel's intake port. The collection array's harmonic dropped a full octave, then another, then faded into silence. The indicator lights on the storage core shifted from urgent amber to a steady, neutral blue.

"Shutdown initiated," Sera called down. "Controlled power-down sequence running. No fail-safe triggered."

The pulling sensation in Lillian's chest — that persistent drain toward the engine — eased and then stopped entirely.

She lowered her tail. The light in her scales dimmed to their normal gentle shimmer. Around her wrists, the silver and lilac bracelets caught a brief, quiet flash as the energy settled.

The Battle at the Cradle rim was over.

But Solvane Criix had not been among the guards.

Lillian knew this because she had been looking for him throughout every second of the fight. She turned slowly, scanning the Cradle, and found exactly what she had expected to find and exactly what she had feared: a single door in the base of the storage core, standing open, leading inside.

He was already in there.

"Bowen," she said quietly. "Stay with the guards. Make sure none of them do anything unfortunate."

"Understood." He pulled a length of polymer binding cord from his tool belt with the air of a man who had packed specifically for this.

Lillian flew toward the open door alone.

Chapter 8: Inside the Core

The interior of the storage core was cathedral-quiet.

Ambient light came from the lithium-alloy walls themselves, which emitted a faint, steady pink-silver luminescence — the passive glow of enormous stored energy now cycling through a controlled release. The air tasted of ozone and something older and more mineral, like the inside of a cave that had never been touched by weather.

The projection matrix occupied the core's center: a towering assembly of crystalline discs arranged in concentric rings on a rotating armature, each disc the diameter of a cartwheel, ground to optical perfection and faintly iridescent. Even in shutdown mode, they turned slowly, their surfaces catching the wall-light and throwing it in slow, sweeping patterns across the floor.

Solvane Criix stood at the armature's base with his back to her.

He was taller than she had expected, well over six feet, thin in the way of someone who had forgotten meals for years in favor of work — his coat hung from his shoulders the way clothes hang from a frame rather than a body — with iron-gray hair pulled back loosely and the posture of someone who had spent the better part of his adult life bending over workbenches. He wore a long coat of deep charcoal, heavily patched, with the remnant insignia of the Prism Institute still visible on the left shoulder — worn, she thought, not as pride but as something between a wound and a reminder.

He spoke without turning. "I felt the shutdown engage." His voice was measured, educated, carrying the cadence of someone who had lectured for years. "Your chemist is good."

"She is," Lillian said, landing near the door.

Criix turned. His face was angular and deeply lined, and his eyes were the particular dark shade of brown that looks almost black in low light. He regarded her with an expression she had not expected: not rage, not defiance, but a tired, searching intensity that reminded her uncomfortably of someone trying to solve a problem they had been working on for too long.

"Lillian the Elemental Dragon," he said. "I've read everything published about lithium magic. Which is not a great deal. You are, as far as the Council's records indicate, unique."

"I'm aware," she said.

"Then you understand what I was trying to do." He said it not as a challenge but as a genuine question.

Lillian let a quiet extension of her awareness settle over the room before answering — a technique she thought of, when she bothered to think of it at all, as a Resonant Read, tuning her senses the way Sera's spectrometer tuned to a specific frequency, listening less for words than for the truth underneath them.

"I understand what you believed you were trying to do," she said carefully. "You believed centralized control of MarBryn's energy infrastructure would create stability. Efficiency. An end to the inequity of energy distribution between the wealthy port cities and the outer regions."

Something shifted in his face. "You did read the papers."

"All of them. Including the ones the Institute suppressed." She took a measured step forward. "Your early research was sound, Criix. The inequity problem is real. The outer regions of MarBryn run on failing old-tech because the investment goes to the ports. People suffer because of that imbalance. You were right about the problem."

"But?" His jaw tightened.

"But a forced seizure of energy infrastructure gives control to one person. You. And no problem of inequity has ever been solved by replacing one concentrated power with another." She held his gaze. "You know this. I think you've known it for a while, and the work became the point rather than the goal."

A long silence stretched between them. The matrix discs turned their slow, reflective arcs.

When Criix finally spoke, his voice was quieter. "It took twelve years. Every component, every calculation, every resource I could acquire. I gave everything to this." He looked at the matrix above him. "And you shut it down in an afternoon."

"We shut down the engine," Lillian said. "Not the problem it was built to address." She paused, choosing her words with the precision she brought to everything. "The Council's field agent who came through that door with me — he's outside right now. The Council sent him to stop the engine, not to erase the underlying research. If you're willing to work within a framework that doesn't require anyone to be connected to a machine against their will, that research has legitimate value."

Criix stared at her. "You're offering me a way forward."

"I'm telling you one exists," she said. "What you do with that is your choice."

He was quiet for a long time, standing in the light of the matrix he had spent twelve years building. Then he reached into his coat, removed a small data-tablet — the kind used for technical documentation, its screen filled with schematics — and held it out toward her.

"The complete engine design," he said. "Every component. The energy distribution architecture. If it were redesigned as a cooperative system instead of a controlled one — power flowing out rather than being extracted — it could do exactly what I intended it to do." He paused. "Without requiring your energy."

Lillian crossed the space between them and took the tablet. She looked at the first schematic for a moment, and her crest-plates rose slowly with the rhythm they had when pieces of a puzzle locked together.

"This will take years to redesign properly," she said.

"Yes," he said.

"And oversight. Independent oversight."

"I know."

She held the tablet and looked at him for one more moment. Then she activated her mind shield — a quiet, thorough extension of it outward, reading the room not for threat but for deception, for the hidden angle, for the concealed intention she might have missed.

She found none. What she found was a man who was exhausted down to his bones and who had, in the span of a short, honest conversation, arrived at something he had perhaps not permitted himself to reach in twelve years of single-minded work: a stopping place.

"Come outside," she said. "Davan Marsh will want to hear all of this, and he'll want to hear it from you directly. And I suspect someone out there can find you an actual meal. You look like you've forgotten what one is."

Something that was almost, but not quite, a laugh escaped him. "It has been a while."

Criix nodded once. He took a last look at the matrix — the life's work turning its slow, luminous circles in the quiet core — and then he followed her out the door.

Chapter 9: What Remains After the Storm

The controlled shutdown of the resonance engine took four more hours to complete fully.

Bowen managed it with Criix's technical guidance and his own considerable competence, working systematically through the engine's power-down sequence while Davan Marsh documented everything on a Council field recorder — a compact device that captured both audio and visual data on a treated alchemical film that could not be altered after recording. Old-world reliability for new-world accountability. True to Lillian's word, one of the guards who had chosen to sit rather than run turned out to have a modest cook-kit in his pack, and Davan ate an entire tin of warmed field stew with the slow, reverent focus of a man rediscovering a lost pleasure, declaring it — despite being, by any objective measure, fairly mediocre stew — the finest meal of his career.

Sera worked alongside Criix, asking questions with the focused, relentless curiosity of someone who has encountered a problem she intends to understand completely. Criix answered with the precision of a man who had been waiting a long time for someone technically equipped to follow his explanations. By the second hour, they were finishing each other's technical sentences in a way that Bowen found mildly alarming and Zix found deeply unsurprising.

"Chemists," Zix said to Lillian, with the air of someone noting that water was wet.

Lillian sat on a low section of the Cradle rim while the work continued, her journal open, the diamond charm at her tail orbiting in its familiar contented loops. She was writing, but not in the crisp observation-notation she used for technical matters. These were longer sentences, more reflective, filled with the particular kind of thought that happens after something difficult has resolved in an unexpected direction.

Davan came and sat beside her after a while, his wire-framed spectacles pushed up on his forehead. He looked tired in the good way — the way of someone who has finished a long, hard job.

"The Council will want a full debrief," he said. "Everything. The engine, Criix, the funding source, the timeline. It'll take weeks."

"I know," Lillian said.

"They'll also want to speak with you." He paused. "You're the reason the shutdown was viable. Your elemental synthesis of the compound component — that's not in any Council protocol. You improvised a solution that their best chemists would have taken days to prepare in a laboratory."

Lillian looked at the vial Sera had returned to her after the shutdown — the tiny measure of lithium carbonate crystal she had produced from her own elemental nature. She rotated it back and forth. "It wasn't improvisation. It was understanding what I am and using it."

"Is there a difference?"

She considered. "Yes. Improvisation is reacting to a situation. Understanding is knowing what you carry and choosing when to use it." She put the vial down. "I've been learning that distinction for years."

Davan was quiet for a moment. "For what it's worth — what you did in there with Criix. Talking to him rather than through him. That was the hardest part of this whole operation, and you made it look straightforward."

"It wasn't straightforward," she said honestly. "I had to decide, in the middle of a fight, that the person I was fighting toward was worth talking to. That's not a calculation. That's a choice."

He nodded slowly. "Well. The Council will note that too."

"Tell them not to make too much of it," she said. "It's just what lithium does. It stabilizes."

From across the Cradle, Bowen's voice rose with the particular tone it got when something worked exactly as intended: "Power-down complete. All arrays inactive. Storage core at zero charge." A pause. "The grid across the Shelf should start restoring within the hour."

Sera looked up from her work with Criix. "The Cinderhollow battery vault will be the first to see restoration. Same signature, reversed."

"Third time this week," Lillian murmured, almost to herself, thinking of the entry she had written in her journal that first morning on the Aqueduct Bridge. Except now the meaning had flipped entirely — from evidence of theft to evidence of return.

She closed her journal and stood.

The four guards who had chosen to remain seated rather than run were secured with Bowen's polymer cord and Davan's field restraints, waiting for Council transport that Davan had already summoned via a long-range signal device concealed in his coat lining. The four who had fled were, as Davan noted pragmatically, someone else's problem for now but not forever. The Council's reach was long and its record-keeping meticulous.

Criix had agreed to return to Dralkethorn voluntarily and face the Council's review. He had not made this agreement with enthusiasm, but he had made it with the quiet resolution of someone choosing a harder and more honest road over a ruined one. He sat apart from the group for the most part, reviewing his own data-tablet with the focused attention of someone already planning the first redesign, though he did, notably, accept a second tin of the guard's mediocre stew when Davan offered it, which everyone present privately understood to be a good sign.

Late in the afternoon, with the chemical fog beginning its evening thickening, Zix landed on the Cradle rim beside Lillian. Her sulfur-yellow eyes were warmer than their first meeting — an impression less of color change than of some interior adjustment.

"The Council will be pleased," Zix said. "They sent me expecting to guide an elemental dragon into a dangerous situation. I didn't expect—" she gestured at the scene below: Bowen packing his tools, Sera filling notebooks, Davan recording, Criix calculating. "—all of this."

"Neither did I," Lillian said.

"You built something in four days," Zix said. "A team. Each person where they needed to be."

"Each person where they needed to be," Lillian corrected gently. "I didn't assign them. I recognized them."

Zix was quiet for a moment. Then, in a different tone: "I'll be returning to the Council with Davan and Criix. My assignment ends here." She looked at Lillian directly. "But I would be glad to work with you again, if the occasion arose."

Lillian smiled — a genuine, unhurried expression that settled comfortably on her features. "I expect occasions will arise. MarBryn has a great deal of work left to do."

As the evening settled into the Hollow and the first battery-grid lights far to the south began to blink back on — visible even at this distance as a faint, warm amber glow above the fog line — Bowen stood beside Lillian and watched them come to life one by one.

"Cinderhollow first," he said with quiet satisfaction.

"Yes," she said.

He was silent for a moment. "What happens now? I mean, for us. For you."

The diamond charm at her tail orbited steadily. Her bracelets shimmered in the last of the day's light — silver and turquoise, lilac and twinkling charm.

"We go home first," she said. "Then we see what the next problem is."

Bowen nodded as if this were the most sensible answer possible, which, coming from Lillian, it was.

Chapter 10: The Lights Come Back On

They left Ashfen Hollow the following morning.

The return through Prism Bog was easier than the entry — Sera's emitter-suppression program had disrupted the detection network thoroughly enough that the path felt almost quiet, the bog's prismatic colors shifting in their unhurried, ancient rhythms without threat or complexity. Zix led them out with the same easy confidence she had led them in, her staff spinning its contented arcs, though she paused at the bog's southern edge to say her farewells.

She gripped Lillian's forearm in the formal Compass Sprite gesture of parting — both hands, briefly, then released. "The Council will have questions," she said. "Be patient with the slow ones."

"I always am," Lillian said.

Davan shook hands with each of them in turn, reaching Lillian last. He held her paw a moment longer than the others. "The work you described in there — the redesigned system. If Criix can be guided toward it properly, it would change MarBryn's energy landscape in a generation." He studied her with his sharp dark eyes. "That would be a significant thing."

"That was always the significant thing," she said. "The engine was just the wrong shape for it."

Criix, standing a little apart with his data-tablet under his arm, met her gaze across the group. He gave a single, careful nod. She returned it.

Then Zix, Davan, and Criix turned northeast toward the Council road to Dralkethorn. Lillian watched them until the fog received them, and then she turned south with Bowen and Sera toward the long climb back up the Verdant Shelf.

The trip home took two days, with a night's rest at a waystation inn called The Copper Compass — a low-beamed, warm-lit establishment halfway up the Shelf road where the proprietor brewed tea from Shelf-grown mineral herbs and asked no questions about mud-stained coats or tool belts or the small, mint-green dragon with luminous scales who took the corner table and spent the evening writing in her journal.

Dinner at the Copper Compass turned out to be the best meal any of them had eaten in four days, and each of them met it according to type. Bowen ordered a meat pie the size of his own head, ate the entire thing along with a second basket of bread, and fell asleep in the chair nearest the fire approximately forty minutes after they arrived, his goggles still on his forehead and his tool belt still buckled, snoring lightly in the particular key of someone with a completely clear conscience. Sera ordered the smallest thing on the menu, a plate of sharp pickled vegetables and hard cheese, and spent the evening at the same table working through her collected data with the focus of someone who would not sleep easily until she understood everything completely.

Lillian asked only for a pot of the proprietor's mineral tea, brewed strong and faintly sweet the way lithia water always was when it came from a good spring, and found it, she told the innkeeper with complete sincerity, the best cup she'd had since Cinderhollow.

Around the second hour, Sera looked up from her spectrometer's display and said, without particular preamble: "I'm going to write this up. The full technical account. Criix's engine design, the elemental synthesis application, the shutdown methodology. It should be in the scientific record."

"It should," Lillian agreed.

"I'll need to interview you properly. For the lithium synthesis section."

"I know."

Sera looked at her for a moment. "You knew that was coming."

"I suspected," Lillian said. "I don't mind. Understanding lithium magic — even partially, even imperfectly — matters for what comes next."

"The redesign," Sera said.

"And whatever else needs doing," Lillian said. "MarBryn runs on energy in every sense of the word. The more people understand how that energy works, the better the decisions they make about it."

Sera made a note. Then she made another note about the first note. Then she closed her spectrometer with the characteristic snap of someone done working for the night, which Lillian recognized as a significant gesture of restraint.

Bowen had fallen asleep in the chair nearest the fire approximately forty minutes after they arrived, his goggles still on his forehead and his tool belt still buckled. He was snoring lightly in the particular key of someone with a completely clear conscience.

Lillian looked at him with the quiet, careful affection she had for very few things.

She thought about what Davan had said — "you built something in four days" — and about her own correction. She had not built anything. She had recognized people: Bowen with his circuit-board mind and his absence of magic and his absolute presence in every situation; Sera with her copper-red hair and her spectrometer and her relentless, clarifying precision; Zix with her sulfur-yellow eyes and her knowledge of paths through difficult terrain; Davan with his fourteen months of patient, solitary work in a fog-country that most people preferred to pretend didn't exist. And Criix — who was neither ally nor enemy at the end, but something more complicated and more important than either.

Each of them had carried something the others needed. That was not strategy. That was, she thought, simply what happened when you paid attention to people.

They reached Cinderhollow on the second afternoon, climbing the last switchback as the sun reached its low, golden, late-day angle that lit the Shelf's limestone face in copper and amber. The village was audible before it was visible: the hiss of gas-lamps, the hum of the battery-grid, the ordinary, beautiful noise of a community going about its evening.

The battery vault lights were green across the board. Every cell charged. Every indicator steady.

Bowen stopped in the middle of the road and looked at the town with his hands on his hips and his goggles pushed up and said nothing for a moment that was not quite silence because it was too full of something.

"It worked," he said finally.

"It worked," Sera confirmed, already noting it.

Lillian flew over to the Aqueduct Bridge — her bridge, the one where this had all begun with a journal entry and a morning's fog — climbing in one last easy pass of Buoyant Lift before settling on the wide stone railing exactly where she had been sitting five days earlier, as though the whole adventure had simply folded back on itself. The marsh below glimmered in the low light. The gas-lamps along the cobblestones came on one by one as the evening deepened, amber and steady and warm. The solar-lattice panels on the rooftops completed their day's work and handed their stored charge smoothly to the grid below, as they were designed to do. Old world and new world, still holding hands, still arguing politely.

Bowen came to stand on the bridge, a fresh twist of honey-grain bread from Marta's already in hand, because some habits simply reasserted themselves the moment a person got home. Then Sera, who looked at the view with the expression of someone noting that it was genuinely beautiful and that this was important data. They stood together in the comfortable quiet that belongs to people who have been through something difficult and come out the other side.

The diamond charm at Lillian's tail orbited in its slow, contented arc. Her bracelets caught the lamp light — a quiet shimmer of silver and lilac. Her scales held their sea-glass glow against the darkening sky.

She opened her journal to a new page and wrote one line:

Cinderhollow. Grid restored. All lights on. Begin next chapter.

She looked up at the lit village, at her two companions on the bridge beside her, at the wide and complicated land of MarBryn spreading out in every direction under a sky that was just beginning to show its first careful stars.

She closed the journal.

The lights of Cinderhollow were bright, and they were staying on.

Lithium Facts

- First detected in 1817 in Stockholm
- More brittle than other alkali metals
- Good conductor of heat and electricity
- Not inherently magnetic
- Ductile and malleable
- Alkali Metal

Author's Note

When I sat down to write Lillian the Elemental Dragon with Lithium Magic, I already knew two things: I wanted her to be small, and I wanted her to be exactly as powerful as lithium actually is.

That's the strange, wonderful contradiction at the heart of Element 3. Lithium is the lightest metal on the entire periodic table — so light it floats on water, so soft you can cut it with a butter knife — and yet it's the element we've built our modern world's batteries around. It stores energy other materials can't hold. It reacts. It moves. It refuses to sit still. Once I understood that, Lillian wrote herself: a compact, sea-glass dragon who looks unassuming right up until the moment she isn't.

This is the third standalone adventure in the Magical Elementals Chemistry Fantasy Series, and by now the shape of the project should be clear — I'm working my way through the periodic table, one element and one elemental at a time. Book One introduced Hildy the Elemental Dragon and hydrogen. Book Two gave you Hetha the Alchemical Wizard and helium. Now we have Lillian the Elemental Dragon and lithium, alkali metal number three, the first true metal in our lineup. There are 118 known elements on that table, which means, yes, I am fully aware of the scale of what I've signed up for. I like to think of it less as a series and more as a very long, very joyful homework assignment I get to hand to myself.

Every book starts the same way for me: research first, story second. Before Lillian ever climbed onto that stone railing over the Aqueduct Bridge, I spent weeks with lithium itself — its place in the alkali metal family, its role in battery technology, its odd chemical restlessness, even its old, almost forgotten history as a supposed health tonic in mineral water. I wanted a middle-grade reader to close this book knowing something true about conductivity and electrochemistry without ever feeling like they'd sat through a lecture. In this expanded edition, I went a step further and gave each of Lillian's abilities its own name, so a reader (or a curious parent) could look up exactly which piece of real lithium science each power is standing on. If Lillian's claws sparking against a copper wire teaches a nine-year-old what a conductor actually does, then the story has done its job twice over — once as adventure, once as science.

I also wanted this book to ask what happens when good science is used the wrong way? Solvane Criix isn't a cartoon villain. He's a brilliant chemist-mage who saw a real problem and chose the wrong solution. I think that's a conversation worth having with young readers early — that ethics and chemistry are never really separate subjects.

Thank you for trusting me with another trip through MarBryn. Whether you've been with this series since hydrogen or you just met Lillian on this stone bridge, I'm glad you're here. There are 115 elements left to go (unless something new is discovered before I finish this series), and I have a feeling some of them are going to surprise us both.

With gratitude and a small electric spark,,

Sybrina Durant

PS: If you're interested in reading the resources the scientific data in this book was derived from, check out this webpage:
https://magicalptelements.com/mdapt/

About the Author

Sybrina Durant is the author of

"The Magical Elementals Chemistry Fantasy Series"

"The Magical Elements of the Periodic Table Series"

"The No Metal No Magic Series"

"The Blue Unicorn – Journey To Osm" books

and she is the creator of the "Elemental Fantasy Periodic Table" which is available as a brilliantly colored poster.

There are also book activities, songs, trading cards and t-shirts for all 118 elements and elementals, too.

Visit MagicalPTElements.com to see everything that is currently available.

If you love to sing and learn, check out the "Chemistry Sing Alongs" playlist on the Sybrinad channel on Youtube.

Glossary of Scientific Terms in Lillian the Elemental Dragon with Lithium Magic

Alchemical Gas

Real Science: A stand-in for real-world refined fuel gases — like natural gas or propane — that burn cleanly to power lighting and heating.

Story Magic: The magically infused gas that lights Cinderhollow's copper-pipe streetlamps and fuels Bowen's camp stove, representing the "old world" energy source the lithium battery-grid is gradually replacing.

Alkali Metal

Real Science: The reactive family of metals in Group 1 of the periodic table — including lithium, sodium, and potassium — known for having a single loosely held outer electron that they give up easily.

Story Magic: Lillian's restless, quick-to-act nature and her electrokinesis are written as echoes of that same eager reactivity.

Arc-Suppressor

Real Science: A real engineering concept: a device that interrupts unwanted electrical arcing or discharge, often using directed electromagnetic interference.

Story Magic: Bowen's handheld invention, built without any magic at all, that disables enemy equipment during the battle at the Cradle — his non-magical answer to what Lillian does instinctively.

Atmospheric Gas

Real Science: The real mixture of gases — mostly nitrogen and oxygen — that make up the air of a planet's atmosphere.

Story Magic: The ordinary MarBryn air that Zix's staff compresses into a spinning globe for her navigation and combat bursts.

Battery-Grid / Electrical Grid

Real Science: A real interconnected network of batteries and power lines that stores and distributes electricity across homes and towns.

Story Magic: Cinderhollow's underground network of lithium cells — the exact thing Criix siphons from, and the thing Lillian is elementally drawn to protect.

Buffer Pulse

Real Science: Lithium carbonate is a real medication used to help stabilize mood in some medical conditions — it acts as a chemical "buffer" that steadies the body's internal chemistry.

Story Magic: Lillian's healing technique, a calm golden current she uses on Davan's cracked ribs — the story's magical version of lithium's real stabilizing, steadying role in the body.

Buoyant Lift

Real Science: Lithium is the lightest metal on the periodic table, with roughly half the density of water — light enough that a chunk of it actually floats.

Story Magic: The name for how Lillian flies: her body becomes unnaturally light for its size, letting a six-foot wingspan carry a roughly forty-pound dragon with startling ease.

Carbonate Bloom

Real Science: Lithium carbonate (Li_2CO_3) is a real, stable chemical compound formed when lithium combines with carbon and oxygen — used in everything from ceramics to medicine.

Story Magic: The technique Lillian uses to condense pure elemental energy into actual crystals of lithium carbonate in a vial, supplying the missing ingredient for the engine's shutdown key.

Carrier Frequency

Real Science: In real electronics, a carrier frequency is the base signal that a device broadcasts on, which other equipment can detect and lock onto.

Story Magic: The energy pattern sustaining Criix's fog wolf constructs, detected by Sera's spectrometer and traced back to a corrupted echo of true lithium resonance.

Cascade Discharge

Real Science: Modeled on a chain reaction or reactive cascade in chemistry, where one triggered reaction sets off a rapid series of others.

Story Magic: Lillian's wide-arc lightning attack that jumps between all six fog wolf emitters at once, disabling them in a single chained burst.

Chemical-Key System

Real Science: Real "lock and key" chemistry: some systems are designed to respond only to a precise compound, resistant to electronic override — a genuine security principle used to prevent remote tampering.

Story Magic: The resonance engine's shutdown panel, which only responds to a specific chemical compound (built from Lillian's own lithium carbonate) rather than to any electrical or magical signal, including Lillian's own.

Collection Array

Real Science: Loosely modeled on real antenna arrays and solar collection grids, which gather ambient energy over a wide area.

Story Magic: The web of copper and silver wire ringing the Sunken Cradle, built specifically to gather — and eventually draw directly from — Lillian's own lithium energy.

Conductivity / Mineral Conductivity

Real Science: A real measurable property describing how easily a material lets electric current pass through it; different minerals and ground compositions conduct differently.

Story Magic: The principle Lillian uses to navigate Prism Bog, sensing the difference between conductive lithium-slate underfoot and non-conductive mud.

Conductor

Real Science: Any real material — usually a metal — that allows electricity to flow through it easily.

Story Magic: Lillian herself, in the moment she presses her claws to a live wire and draws its stored current into her own body, the way a conductor grounds a charge.

Current / Current Signature

Real Science: Real electric current is the flow of charge through a conductor; sensitive instruments can detect and distinguish patterns, or "signatures," in that flow.

Story Magic: The unique energy fingerprint Lillian reads with her claws, letting her identify a deliberate power drain and later match it to Criix himself.

Current Analyzer

Real Science: A real instrument used by electricians and engineers to measure the strength and direction of electrical current.

Story Magic: Bowen's tool, the mechanical, non-magical counterpart to Lillian's innate ability to sense current signatures by touch.

Electrochemical

Real Science: A real branch of chemistry studying the relationship between electricity and chemical reactions — the basis of how every battery works.

Story Magic: Both Criix's siphoning devices and Lillian's own elemental nature are described as "electrochemical," tying the villain's technology and the heroine's magic to the same real science.

Electrokinesis

Real Science: Not a real, scientifically verified ability — but inspired by how real electricity moves through conductive material.

Story Magic: Lillian's core magical power: generating, directing, and controlling electrical current at will, used for everything from lightning bolts to shields.

Electromagnetic Field / Resonance Field

Real Science: A real field of force generated by electrically charged objects and magnets, capable of interacting with matter and other energy at a distance.

Story Magic: The field Lillian senses beneath Prism Bog and around the resonance engine, which is deliberately tuned to resonate with — and pull on — her own lithium energy.

Elemental Lithium / Lithium Magic

Real Science: Real elemental lithium (Li) is a soft, silvery, highly reactive alkali metal, element number 3 on the periodic table.

Story Magic: The living, biological version of that element's power that runs through Lillian's body — the one thing in the story that no machine, including Criix's engine, can manufacture on its own.

Full Discharge

Real Science: A real event in battery science: releasing all of a battery's stored energy at once, either intentionally or as an uncontrolled failure.

Story Magic: Lillian's final, deliberate release of nearly all her stored energy in one wide, non-destructive electromagnetic pulse that disables every piece of enemy equipment in the Cradle.

Galvanically Isolated

Real Science: A real electrical engineering term for a system deliberately disconnected from outside circuits so no stray current or signal can cross into it.

Story Magic: The reason the resonance engine's shutdown panel can't simply be switched off with electricity or magic — it has to be triggered with a physical chemical key instead.

Holographic Scroll

Real Science: Inspired by real photochemical and digital display technology, where treated film or screens can be made to project or display text and images.

Story Magic: The alchemically treated film Lillian reads in the Cinderhollow archive, glowing with soft blue light when pressed to a reading panel.

Ion Lance

Real Science: Lithium has the lowest first ionization energy of almost any metal, meaning it gives up its single outer electron more easily than most elements — a real, measurable chemical property.

Story Magic: Lillian's sharp, targeted lightning bolt, fired from her clawtips — a single fast release of energy standing in for that easy electron transfer.

Ionic Ward

Real Science: Real ionic bonds form tight, highly stable lattices, like the strong crystal structure of table salt — a genuine source of physical strength in chemistry.

Story Magic: The broad, shimmering energy barrier Lillian holds between attackers and her friends, absorbing charge bolts the way a tightly bonded ionic lattice locks ions firmly in place.

Lattice Grip

Real Science: Real lithium metal has an orderly crystal lattice structure that gives it physical strength and shape despite being soft and light.

Story Magic: Lillian's telekinetic ability to lift, redirect, and hold objects with focused will, imagined as reaching into that same kind of invisible structural order.

Lithia Water

Real Science: A real historical product — natural or manufactured lithium-enriched mineral water was sold in the late 1800s and early 1900s and marketed as a health tonic (lithium citrate was even an original ingredient in 7-Up).

Story Magic: Lillian's preferred drink, a mineral-rich spring water she says "settles her thinking" after using her magic.

Lithium-Alloy / Lithium-Slate

Real Science: Real lithium alloys and lithium-bearing minerals (like spodumene) are mined and refined for use in batteries, ceramics, and glass.

Story Magic: The metal of the battery vault's cells and the resonance engine's storage core, and the mineral deposits beneath Prism Bog that Lillian can feel through her feet.

Lithium Carbonate

Real Science: A real, stable chemical compound (Li2CO3) refined from lithium ore, used industrially in glass and ceramics and medically as a mood stabilizer.

Story Magic: The compound Lillian synthesizes directly from her own body in Chapter Six — the clearest proof in the story that her power is chemical, not just electrical.

Magnetism / Magnetic

Real Science: A real physical force generated by moving electric charges, capable of attraction and repulsion.

Story Magic: The force behind the resonance engine's collection array, and behind Lillian's own final disruption pulse, which floods the area with her field to short out enemy equipment.

Malleable & Ductile

Real Science: Real physical properties of many metals, including lithium, that allow them to be hammered into shapes or drawn into wire without snapping.

Story Magic: A description of how flexibly Lillian can shape her own magic — from a sharp bolt to a broad shield — echoing the real adaptability of the metal she's named for.

Projection Matrix

Real Science: Loosely modeled on real optical systems that use arrays of lenses or crystals to focus and direct energy or light.

Story Magic: The rotating assembly of crystalline discs inside the storage core that would have focused Criix's stolen energy into a continent-wide broadcast.

Resonance Engine

Real Science: Inspired by the real physics idea of resonance, where a system responds strongly to energy at its own natural frequency.

Story Magic: Criix's machine, purpose-built to resonate with — and forcibly draw out — living elemental lithium energy, meaning Lillian's energy specifically.

Resonant Read

Real Science: Inspired by real spectrometry, where instruments are tuned to detect a very specific frequency or signature within a larger signal.

Story Magic: Lillian's quiet extension of awareness in Chapter Eight, used to sense whether Criix is telling the truth rather than to sense electricity.

Solar-Lattice Panels

Real Science: Real photovoltaic solar panels use a lattice of crystalline silicon cells to convert sunlight directly into electricity.

Story Magic: Cinderhollow's rooftop panels, which feed sunlight into the town's lithium battery-grid every day, and are shown restoring power alongside Lillian's efforts at the story's end.

Spectrometer

Real Science: A real scientific instrument that identifies chemical compounds and measures energy across different frequencies of light or radiation.

Story Magic: Sera's field instrument, which first detects Lillian's elemental signature and later maps the fog wolves' carrier frequency.

Storage Core

Real Science: Modeled on real large-scale battery storage systems, which hold accumulated electrical energy in stable form until it's needed.

Story Magic: The house-sized structure at the center of the Sunken Cradle, built from lithium-alloy plating and glowing the same pink-silver as real metallic lithium.

Telekinesis

Real Science: Not a real, scientifically verified ability.

Story Magic: Lillian's secondary power, letting her move objects with focused will — framed in the story as a physical, kinetic expression of her elemental nature rather than a purely electrical one.

Thermal-Coil

Real Science: A real heating element, like the coil in an electric stove or space heater, that converts electrical energy into heat through resistance.

Story Magic: The heating element in Bowen's folding camp stove, a small human-engineered echo of the much larger energy transformations Lillian performs constantly.

Voltaic Draw

Real Science: Named for the real voltaic (galvanic) cell, the basic unit of a battery, in which a reactive metal like lithium gives up electrons to generate current.

Story Magic: Lillian's technique for drawing stored current out of an object and into her own body, essentially running a battery's basic process in reverse.

www.ingramcontent.com/pod-product-compliance
Lightning Source LLC
LaVergne TN
LVHW050827061126
851857LV00012B/1024
9781942740728